Food Innovation and Engineering (FOODIE) Conference 2019

Philadelphia, Pennsylvania, USA
8 – 10 December 2019

ISBN: 978-1-7138-0998-2

Printed from e-media with permission by:

Curran Associates, Inc.
57 Morehouse Lane
Red Hook, NY 12571

Some format issues inherent in the e-media version may also appear in this print version.

Printed with permission by Curran Associates, Inc. (2020)

For permission requests, please contact AIChE
at the address below.

AIChE
120 Wall Street, FL 23
New York, NY 10005-4020

Phone: (800) 242-4363
Fax: (203) 775-5177

www.aiche.org

Additional copies of this publication are available from:

Curran Associates, Inc.
57 Morehouse Lane
Red Hook, NY 12571 USA
Phone: 845-758-0400
Fax: 845-758-2633
Email: curran@proceedings.com
Web: www.proceedings.com

TABLE OF CONTENTS

KEYNOTE

HEALTH & FOOD SAFETY

PROCESSING

KEYNOTE 2

PROCESSING (CONT)

FOOD ARCHITECTURE

SUSTAINABILTY & SECURITY

The Intersection of Technology & Real Food

FOODIE
FOOD INNOVATION AND ENGINEERING CONFERENCE

Conference: Food Innovation and Engineering Conference

Year: 2019

Proceeding: 2019 Food Innovation and Engineering (FOODIE) Conference

Group: General Submissions

Session: Keynote 1

Time: Monday, December 9, 2019 - 8:00am-8:45am

Authors:

Slavtcheff, C., Campbell's Soup Company

Consumers have long demanded 5 simple things in the food they buy and consume. It has to be safe, nutritious, delicious, accessible and affordable. The modern day food system has been built around 150 years of advances in science to provide an ever higher degree of delivery against these benefits. In the past decade, however, consumers have fundamentally shifted their purchasing criteria around their perception of â€œsafeâ€, providing an ever higher degree of scrutiny on what is in their food and how it is grown and processed. The food industry is now challenged with maintaining the upward trajectory of delivery against those five benefits, but with an ingredient toolbox that is limited to ingredients that are recognizable and desirable.

3D Time Variant Reactor Engineering Model for Red Wine Fermentations

Conference: Food Innovation and Engineering Conference

Year: 2019

Proceeding: 2019 Food Innovation and Engineering (FOODIE) Conference

Group: General Submissions

Session: Health and Food Safety

Time: Monday, December 9, 2019 - 11:15am-11:45am

Authors:

Miller, K., UC Davis
Block, D. E., University of California, Davis

Red wine fermentations are performed in the presence of grape solids (skins and seeds) to ensure extraction phenolics. This results in a heterogeneous fermentation system, as the solids float to the top to form a porous solid-liquid â€œcap.â€ Modeling of red wine fermentations is, therefore, complex and must consider spatial heterogeneity to predict fermentation kinetics. We have developed a reactor engineering model for red wine fermentations that includes the fundamentals of fermentation kinetics, heat transfer, diffusion, and compressible fluid flow. COMSOL was used to solve all components of the model simultaneously utilizing a Finite Elements Analysis (FEA) approach.

A kinetic model for the extraction of phenolics from grapes as a function of temperature and ethanol concentration was derived, incorporating the release of phenolics from grape cells, the adsorption equilibria of phenolics onto cell wall material, and the reaction of phenolics in solution. The findings of this model were supported by bright field and epifluorescence microscopy.

This extraction model was then combined with the fermentation reactor engineering model, enabling prediction of phenolic extraction as a function of fermentation progression and winemaking conditions.

Data-Indicating Transfer of Antimicrobial-Resistance Genes from Food to Humans

Conference: Food Innovation and Engineering Conference

Year: 2019

Proceeding: 2019 Food Innovation and Engineering (FOODIE) Conference

Group: General Submissions

Session: Health and Food Safety

Time: Monday, December 9, 2019 - 12:15pm-12:45pm

Authors:

Hua, M., Germantown Friends School
Huang, W., Central Bucks High School South
Chen, A., The Episcopal Academy
Rehmet, M., Central Bucks High School South
Jin, C., Central Bucks High School South
Huang, Z., Villanova University

Foodborne pathogens cause hundreds of illnesses and become an urgent issue for food safety, as those pathogens are becoming increasingly resistant to commonly used antimicrobials. These pathogens, equipped by antimicrobial resistance (AMR) genes, can be passed from livestock to humans. It is thus important to investigate the trend of the gene transfer between livestock and humans over time. Fortunately, the NCBI Pathogen Detection Isolates Browser Database contains the AMR gene information for isolates from both clinical and environment settings. This work conducts the first multivariate statistical analysis of the antimicrobial genes and pathogens detected in the clinical and environmental samples in the US from 2013 to 2018. The data analysis indicated that antimicrobial-resistance usually begins in environmental settings and spreads to humans through contact with infected meat, specifically poultry. It was also found that *Acinetobacter baumannii*, *E. Coli* and *Shigella*, *Klebsiella pneumoniae*, and *Salmonella enterica* were the most resistant pathogens with *Salmonella enterica* and *E. Coli* and*Shigella* being the main spreaders of resistance genes. Additionally, the genes *fosA*, *oqxB*, *ble*, *floR*, *fosA7*, *mcr-9.1*, *aadA1*, *aadA2*, *ant(2â€™â€™)-Ia*, *aph(3â€™â*

€™)-Ib, aph(3â€™)-Ia, aph(6)-Id, blaTEM-1,qacEdelta1, sul1, sul2, tet(A),
and *tet(B)* exhibited the most antimicrobial-resistance. On the other hand, tetracycline, ceftriaxone, streptomycin, and aztreonam were the antimicrobials to which both foodborne and clinical pathogens were most resistant to. The results of this study may lead to improved foodborne illness prevention regulations and policies that help lower antimicrobial-resistance in pathogens.

Dietary Fibers and Foundation Guild for a Healthy Gut Microbiota

FOODIE
FOOD INNOVATION AND ENGINEERING CONFERENCE

Conference: Food Innovation and Engineering Conference

Year: 2019

Proceeding: 2019 Food Innovation and Engineering (FOODIE) Conference

Group: General Submissions

Session: Health and Food Safety

Time: Monday, December 9, 2019 - 9:45am-10:15am

Authors:

Zhao, L., Rutgers University

Short chain fatty acids (SCFAs) from bacterial fermentation of dietary fibers play essential roles in human nutrition, immunity and metabolism. Our ancestors had much higher intake of dietary fibers than us today. Reduced intake of dietary fibers and diminished diversity of SCFA-producing bacteria may underlie many chronic diseases such as type 2 diabetes and cancer. In patients with type 2 diabetes, increased intake of diverse dietary fibers can selectively promote a group of acetic and butyric acid producers, that can increase insulin secretion by promoting more GLP-1 production in the gut; more importantly, they can acidify the gut environment, produce antimicrobials and occupy available niches to inhibit pathogenic and detrimental bacteria, which may promote inflammation and suppress GLP-1 production. Thus, in addition to providing SCFAs to directly benefit the hosts, this group of SCFA producers plays important ecological functions in the gut microbiota. Akin to tall trees in a closed forest, they work as the â €œfoundation guildâ€ for structuring the healthy gut microbiota. To help patients regain a healthy gut microbiota, this foundation guild must be re-seeded and re-established first.

Not food, not medicine: The limbo world of nutraceuticals

FOODIE
FOOD INNOVATION AND ENGINEERING CONFERENCE

Conference: Food Innovation and Engineering Conference

Year: 2019

Proceeding: 2019 Food Innovation and Engineering (FOODIE) Conference

Group: General Submissions

Session: Health and Food Safety

Time: Monday, December 9, 2019 - 10:15am-10:45am

Authors:

Relkin, N., Brilliant Health LLC

Modern day nutraceuticals occupy a muddled conceptual and regulatory space between foods and medicines. While some are clearly intended to have medicinal effects, nutraceuticals are rarely tested with the same rigor as pharmaceuticals or proven to be as effective for disease treatment. Other nutraceuticals aspire to be â€œsuper-foods,â€ allegedly providing missing elements of nutrition that are not readily delivered through the consumption of regular diets. The safety and effectiveness of such products have not been well-established and few can be said to truly move the needle on disease prevention or other health and wellness measures. For these and other reasons, nutraceuticals have failed to inspire consumer confidence and are treated with relative disdain by the general medical and scientific communities. Ironically, these negative perceptions have not prevented the nutraceutical industry from generating a rapidly growing, multi-billion-dollar per-annum market. This suggests that there is a very real but unmet need for more effective health and wellness products. Fortunately, new conceptual approaches and technological advances have created a potential turning point in the evolution of nutraceuticals. In particular, enhanced analytic and manufacturing methods that foster ensemble effects through the standardization of multiple bioactive constituents in complex food extracts have the potential to capture the health benefits associated with foods while delivering the disease-altering effects of pharmaceuticals. Nutraceuticals created by this approach can alter how we treat and prevent human diseases in unique and meaningful ways but face an uphill battle in terms of commercialization owing to the many false claims and regulatory hurdles that exist in the current nutraceutical marketplace.

Novel Soft Pretzel Snack for Delivery of Soy and Safflower Oil As a Means of Health Promotion

Conference: Food Innovation and Engineering Conference

Year: 2019

Proceeding: 2019 Food Innovation and Engineering (FOODIE) Conference

Group: General Submissions

Session: Health and Food Safety

Time: Monday, December 9, 2019 - 10:45am-11:15am

Authors:

Vodovotz, Y., The Ohio State University
Ahn-Jarvis, J., Quadram Institute
Sosh, D., The Ohio State University
Riedl, K., The Ohio State University
Belury, M., The Ohio State University

In the US, 69% of adults are considered to have excessive body weight that has been associated with a sedentary lifestyle, fast food consumption, and snacking. A functional food, deliberately designed having specific physicochemical properties is an innovative approach to utilizing food ingredients intended for promoting health. Both safflower oil and soy have demonstrated anti-diabetic, pro-thermogenic, and pro-satiety activity in mechanistic and rodent studies. We hypothesize that a functional snack food such as a soft pretzel containing safflower oil either with or without soy compared to the control pretzel (wheat pretzel with no safflower oil) will reduce postprandial blood triglycerides and glucose response. Therefore, our objectives were to 1. evaluate the effects of functional snack foods containing safflower oil with and without soy on endpoints of energy metabolism in postmenopausal women with metabolic syndrome and 2. monitor postprandial satiety using visual analogue scale. Eighteen overweight (BMI > 25 kg/m2) postmenopausal women with metabolic syndrome underwent three 6 hour oral triglyceride tolerance test over 14 weeks. Women had a run-in phase with a control pretzel and then randomized to either the wheat or soy pretzel with safflower oil and then crossover to the other pretzel. Biomarkers of energy metabolism, anthropometric measurements and satiety were assessed. Results indicated no statistical differences in fasting glucose, triglyceride or cholesterol. However, significant differences were observed in serum glucose AUCâ€™s between the soy and wheat pretzel interventions as well as triglyceride AUC between safflower oil and control pretzels. Importantly, the satiety measures showed soy-safflower pretzel suppressed hunger longer than wheat pretzel. This pilot study confirms the potential of a soy-safflower pretzel as an alternative, nutrient dense snack for curbing appetite.

Reducing Foodborne Bacterial Biofilms on Produce and Packaging with an Engineered Enzyme

FOODIE
FOOD INNOVATION AND ENGINEERING CONFERENCE

Conference: Food Innovation and Engineering Conference

Year: 2019

Proceeding: 2019 Food Innovation and Engineering (FOODIE) Conference

Group: General Submissions

Session: Health and Food Safety

Time: Monday, December 9, 2019 - 11:45am-12:15pm

Authors:

Mayton, H., University of Virginia
Berger, B., Lehigh University
Walker, S. L., University of California, Riverside

Biofilm formation is a major cause of post-harvest microbial persistence on fresh produce and packaging surfaces, as the biofilm matrix can provide pathogens and other bacteria with protection from common disinfection approaches. The potential of an enzyme-based food processing aide has been investigated as a green, targeted supplement to common disinfection practices for preventing bacterial adhesion and removing mature biofilms. Crystal violet staining of biofilms formed in multi-well polystyrene plates has been used to demonstrate the efficacy of enzymatic biofilm prevention and removal on *E. coli* O157:H7, *E. coli*25922, *Salmonella* Typhimurium, and *Listeria monocytogenes*. In the presence of 0.1 mg/mL enzyme, biofilm development was significantly inhibited for all bacteria, with a maximum of 41 ± 7% reduction for *E. coli*O157:H7. Effectiveness of mature biofilm removal varied by bacteria species type, with a maximum of 35 ± 12% reduction for *E. coli* O157:H7. A microfluidic flow cell was used to directly observe and quantify the impact of enzyme rinses on *E. coli* O157:H7 cells adhered to spinach leaf surfaces. In the flow cell, enzyme rinses resulted in significantly greater cell removal than water, representing a reversal of initial phases of biofilm formation.

Initial biochemical characterization of the enzyme indicates a key role in surfacial extracellular polysaccharide degradation, suggesting multiple potential mechanisms of biofilm disruption. Transmission electron microscopy (TEM) of treated and untreated cells reveal major modification to surface polysaccharide structures, consistent with the predicted role as a polysaccharide-degrading enzyme. Additionally, changes in cell surface hydrophobicity were measured using the microbial adhesion to hydrocarbons (MATH) test and provide further evidence of changes to the cell surface. These results present a strong case for further development and optimization of enzyme activity for applications as a novel food processing and packaging aid to minimize food safety risks from bacterial pathogens.

The Science of Deliciousness: How Do We Measure Yummyness?

Conference: Food Innovation and Engineering Conference

Year: 2019

Proceeding: 2019 Food Innovation and Engineering (FOODIE) Conference

Group: General Submissions

Session: Health and Food Safety

Time: Monday, December 9, 2019 - 9:15am-9:45am

Authors:

Hopfer, H., The Pennsylvania State University

Sensory Science, the scientific discipline that utilizes the human senses of seeing, hearing, touching, tasting, and smelling to evaluate consumer products, is typically associated with applications in the food industry. However, humans use the same perceptional cues to evaluate non-food products, from personal care products and cosmetics to household appliances, automotives and adhesives. In all these non-food applications, an end-consumer selects products not just based on functionality and product safety, but also on affectional and perceptional cues, such as visual appearance, touch, and smell.

The sensory and consumer science toolbox includes different frameworks and methodologies aimed at understanding consumers' perception and acceptability towards a product. A thorough understanding of a product's appeal to a consumer does not stop at main functional components, but requires a holistic view on the sensory properties, presenting an analytical challenge that requires an interdisciplinary approach, combining instrumental methods with human sensory methodology â€“ after all, the final appeal of flavor, haptics, etc. are judged by a human, not a machine.

Using real-world examples of both food and non-food products, my talk will focus on these major areas: (i) how human sensory perception works, with a focus on smell, taste, and texture; (ii) general human sensory and combined sensory-instrumental approaches to understand the sensory and affective appeal of products, and (iii) my own personal thoughts on sensory and consumer science-assisted product development, based on my own experiences with odor-active components in polyolefin materials, and complex food matrices, such as chocolate and wine.

Applying Multiphysics Modeling and Simulation to Optimize Coating Process in Confectionery

Conference: Food Innovation and Engineering Conference

Year: 2019

Proceeding: 2019 Food Innovation and Engineering (FOODIE) Conference

Group: General Submissions

Session: Processing

Time: Monday, December 9, 2019 - 1:30pm-2:00pm

Authors:

Taye, E., MARS WRIGLEY Confectionery
Mummudi, M., Tridiagonal Solutions Inc.
Shirsath, S., Tridiagonal Solutions

A mathematical method has been developed to model and simulate the candy coating process. The framework of the model has been validated experimentally and further used to conduct DOE to study and identify the optimum process conditions as well as physical phenomena that cannot be detected experimentally. To capture accurately the complex physics of sugar coating we used Discrete Element Modeling (DEM) and Computational Fluid Dynamics (CFD) simulation tools. In this work, we discuss the simulation of factory batch coating process as closely as possible. In the model, the shape of the candy, the coating drum inside structure, process and boundary conditions such as temperature, coater speed, humidity, pressure, air flow, number of spray nozzles, etc. are considered. The coating drum is loaded initially with 1700Kg candy centers and at the end of coating process it will be about 2300Kg. It takes about an average of 5hrs to fully coat candies. The goal of the optimization process is to reduce the total coating cycle time without sacrificing the quality of coating. Based on the DOE simulation we identified the best optimum process parameters, predicted the coating thickness distribution, moisture content distribution in candies and crystal growth rates. The validation of the model in one of factory will be discussed also in this paper.

Considerations for a Successful Regulatory Roadmap for Novel Food Ingredients

Conference: Food Innovation and Engineering Conference

Year: 2019

Proceeding: 2019 Food Innovation and Engineering (FOODIE) Conference

Group: General Submissions

Session: Processing

Time: Monday, December 9, 2019 - 2:30pm-3:00pm

Authors:

Lewis, E., NutraSteward

In developing new food ingredients for market, it is essential to consider the regulatory requirements to gaining market access at the earliest time point. An overview of the regulatory routes to market will be provided and the key factors to include in a drawing up a regulatory roadmap discussed. These factors include but are not be limited to, the potential hurdles, resources, time frames and costs associated with gaining regulatory acceptance. Acknowledging the global nature of marketing strategies, the presentation will focus mainly on the U.S. but also consider the EU and Canada.

From Pressure to Plasma: Non-Thermal Food Processing Technologies Research at the USDA's Eastern Regional Research's Center

FOOD INNOVATION AND ENGINEERING CONFERENCE

Conference: Food Innovation and Engineering Conference

Year: 2019

Proceeding: 2019 Food Innovation and Engineering (FOODIE) Conference

Group: General Submissions

Session: Processing

Time: Monday, December 9, 2019 - 1:00pm-1:30pm

Authors:

Niemira, B. A., USDA-ARS

Foodborne pathogens cause millions of illnesses every year. At the US Department of Agricultureâ€™s Eastern Regional Research Center near Philadelphia, PA, scientists and engineers have focused on developing new ways to improve food safety and shelf life while retaining quality and nutritional value. A variety of technologies have been advanced in recent years to expand on the suite of tools available to food processors. From long-standing research projects on nonthermal processing technologies such as irradiation, high pressure and advanced antimicrobial compounds, to newer approaches such as cold plasma, high intensity light treatments, radio frequency pasteurization, and active packaging, the food safety research at ERRC has addressed key aspects of efficacy, scalability and practicality. This presentation will give an overview of this research, and the drivers for these technologies, such as the desire for minimal processing, clean labels, lower costs, and sustainability.

Mixing the Ingredients for a Fully Baked Asset Care Strategy

Conference: Food Innovation and Engineering Conference

Year: 2019

Proceeding: 2019 Food Innovation and Engineering (FOODIE) Conference

Group: General Submissions

Session: Processing

Time: Monday, December 9, 2019 - 2:00pm-2:30pm

Authors:

Rosenthal, D., Reliability Strategy and Implementation Consultancy LLC

Ensuring the alignment of asset (equipment) care with the demands of manufacturing in the process industries is a challenge for every owner. Achieving uptime, safety and cost goals are vital for a reliable supply chain and competitive business. Asset care professionals that design predictive, preventative and protective strategies for assets will claim that these tasks are identical based on the failure modes for common equipment likes pumps, vessels, piping, etc. However the food processing industry has unique assets starting from food preparation to packaging.

Many of the assets like mixers, cookers, fillers, box makers, tapers, palletizers and wrappers have failure modes that have a short time frame from the early signs of failure to the complete loss of function. Asset care strategies must account for these failure modes and be able to detect and prevent this loss of function. These losses of function place stress on the facility productivity and operating costs.

The key is to employ both Maintenance and Production personnel in looking for the early signs of failure. This involves a redesign of preventative tasks to produce predictive tasks looking for the early signs of failure. In addition, sanitation tasks must be designed not to destroy electrical and mechanical components.

This presentation will provide the right mix of asset care tasks for food processing equipment. These tasks focus on condition-based (CBM) and not time-based tasks. CBM tasks look for the early signs of failure without stopping machine function like preventative tasks. Sanitation practices will be included that protect electrical and mechanical components. Companies that deploy these strategies will be shown to improve safety, overall equipment effectiveness, and production costs. Reliable assets improve safety records and produce predictable supply chains. They provide for the owner a competitive advantage through reduced maintenance costs and improved uptime.

Novel Food Ingredients: Challenges and Opportunities in Regulating an Emerging Industry

Conference: Food Innovation and Engineering Conference

Year: 2019

Proceeding: 2019 Food Innovation and Engineering (FOODIE) Conference

Group: General Submissions

Session: Processing

Time: Monday, December 9, 2019 - 3:00pm-3:30pm

Authors:

Glass, D., D. Glass Associates, Inc.

Advances in biotechnology, cell culture techniques and other innovative technologies are leading the way to the development of a variety of novel foods and new ways of producing and processing food for human and animal consumption. As is often the case with emerging technologies and new industry sectors, questions have been raised about how well government regulatory systems in the U.S. and elsewhere in the world are equipped to efficiently oversee these new products in a manner that ensures product safety and consumer confidence while not overburdening or overly-delaying the introduction of novel products.

This presentation will provide an overview of U.S. and international food and animal feed regulation. The talk will examine how novel foods and food ingredients are currently assessed under these regulatory systems, and how successfully these regulations have been able to address foods produced by traditional biotechnology processes. Case studies will be presented to review how regulations may affect new food products, including the uncertainty and controversy that have arisen regarding U.S. regulatory jurisdiction over meat products produced by cellular agriculture, as well as other novel technologies including the use of GMO plants in food, the use of microbial biomass in animal feed, and the introduction of new dietary supplements, including those derived from hemp or cannabis.

The need for existing regulatory programs to adapt to address new technologies often creates uncertainty among industry and other stakeholders. On the other hand, this also presents an opportunity for all stakeholders to participate in the development of rational, science-based policies that allow novel products to come to market under conditions where consumers are ensured of safety.

Deliciousness Beyond Umami

Conference: Food Innovation and Engineering Conference

Year: 2019

Proceeding: 2019 Food Innovation and Engineering (FOODIE) Conference

Group: General Submissions

Session: Keynote 2

Time: Tuesday, December 10, 2019 - 8:15am-9:00am

Authors:

Eckert, D. M. A., Takasago International Corp.

Our global discovery studies with consumers, food experts and chefs showed that deliciousness is about finding the right combination of flavors, preparing them well to achieve the perfect balance. This definition takes us to what we know from traditional Japanese cuisine. Since Japanese cuisine â€œWashokuâ€ was added to UNESCO Intangible Cultural Heritage list, it has drawn a great deal of attention for being healthy and delicious. Like umami (the fifth taste), one of the feature expression of Japanese flavors and cuisine on a daily basis is the concept of koku. It has been described as being flavorful, rich, full-bodied and well-balanced.

Our technical studies advanced our understanding of the koku concept and enabled the development of a series of unique koku flavor systems. Takasagoâ€™s research shows koku is the overall taste experience from the harmonious integration of aroma, taste, mouthfeel and chemesthesis in foods resulting in rich, full, well-balanced flavor and lingering pleasant aftertaste. Iterative creative flavor development is required to optimize the koku flavor experience.

Takasago initiated a series of studies to explain the Japanese concept of koku for the global flavor market. This presentation will show how we confirmed the relationship of koku and its contribution to intensifying aroma, taste and chemesthesis perception through sensory evaluation in a variety of food and beverage model systems and language validation with consumers. Perception of taste, smell, texture, appearance, temperature, sound, etc., are contributing factors of deliciousness, whereas, the koku concept is complicated and confined to the harmonious sensation in the oral cavity. Koku has temporal properties from initial sensation of flavorful harmony to full body and lingering pleasant aftertaste.

Conformable Surface Dielectric Barrier Discharge Plasma Treatment of Fresh Produce: Evaluation of Microbial Inactivation Efficacy and Quality Attributes

FOODIE
FOOD INNOVATION AND ENGINEERING CONFERENCE

Conference: Food Innovation and Engineering Conference

Year: 2019

Proceeding: 2019 Food Innovation and Engineering (FOODIE) Conference

Group: General Submissions

Session: Processing (cont)

Time: Tuesday, December 10, 2019 - 10:45am-11:00am

Authors:

Salvi, D., North Carolina State University
Wang, Q., North Carolina State University
Pal, R., Rutgers University
Mazzeo, A., Rutgers University

Conformable plasma electrodes offer the advantages of being bendable, lightweight and has the potential for high local concentrations of plasma to interact with exterior surfaces being treated, which are expected to benefit future applications in the food industry. In this study, plasma generated using conformable surface dielectric barrier discharge (SDBD) electrodes made by metallized paper was evaluated for its antimicrobial activity and effects on quality attributes of fresh tomato and baby spinach. Two conformable SDBD systems were prepared: a two-cone configuration of electrodes was used for plasma treatment of tomatoes while the flat circular design was used for spinach leaves. Following plasma treatment for 2 min, 5 min, and 10 min, reductions of 2.7±0.6 log CFU/tomato, 3.9±0.3 log CFU/tomato, and 5.2±0.2 log CFU/tomato of *E. coli* population were achieved on tomato surface, while 2.5±0.1 log CFU/leaf, 2.7±0.9 log CFU/leaf, and 2.9±0.9 log CFU/leaf reduction was achieved on the surface of baby spinach. The firmness of the treated tomato and spinach, evaluated using puncture analysis, revealed no significant difference ($p > 0.05$) in either spinach or tomato treated for 2 min, 5 min, or 10 min compared with untreated control. While no changes in color and visual appearance were observed in treated tomatoes, regional darkening of the spinach leaves was observed. The results demonstrate that the flexible plasma SDBD electrodes are effective as a potential decontamination technique on the food surface. The variation in inactivation between tomato and spinach indicates that the conformability of the SDBD electrodes and the surface properties of food play an important role in affecting the decontamination efficacy. In addition, optimization of the flexible plasma SDBD is needed to maximize the reactive species concentration to achieve microbial inactivation while preserving the quality attributes of treated food.

High Pressure Jet Processing for Clean-Label Foods and Ingredients

Conference: Food Innovation and Engineering Conference

Year: 2019

Proceeding: 2019 Food Innovation and Engineering (FOODIE) Conference

Group: General Submissions

Session: Processing (cont)

Time: Tuesday, December 10, 2019 - 10:30am-10:45am

Authors:

Harte, F., Pennsylvania State University

High Pressure Jet (HPJ) is a nonthermal technology that promotes novel functionality by subjecting liquid foods to hydrostatic pressure (up to 500 MPa), high shear (speeds up to Mach 4) and impingement. A liquid is pressurized by an intensifier pump and then a restriction to flow (diamond nozzle) creates an aerosol that is immediately recovered in a tube-in tube heat exchanger. Experiments conducted in liquid milk, demonstrated that extensive casein micelle dissociation occurs at processing pressure >350 MPa resulting in increased viscosity (3.9 mPa.s*vs.* 1.2 mPa.s in control), foaming (foam expansion index, FEI= 116% *vs.*63% in controls) and emulsifying (emulsion activity index, EAI=8.3% *vs.*6.1% in controls) of both skim and whole milk. Additionally, ultracentrifugation studies showed a strong interaction between milk fat triglycerides and casein proteins, resulting in co-precipitation of protein and fat after HPJ processing. In a different configuration, the HPJ was connected to a modified spray dryer (SD) for the direct drying of HPJ processed liquid streams. The reconstituted powders showed improved foaming (FEI=114% *vs.* 60% in controls) and emulsifying (ESI=148% *vs.*134%) properties indicating that the transformation to powder ingredients did not negatively affect the observed functionality. In further experiments, protein-protein and protein-fat interactions allowed for the stabilization of a protein-rich, carrageenan-free, HPJ chocolate milk for 14 day without appreciable phase separation. The use of HPJ also showed potential for the stabilization of plant proteins (e.g., soy protein) in liquid foods, and the formation of stable foams in concentrated protein systems (e.g., ice cream). HPJ technology has a strong potential as a processing operation for the development of novel clean label foods and food ingredients.

Importance of Measuring the Biomass Composition of Saccharomyces Cerevisiae to Model Wine Fermentations Using a Genome-Scale Model

Conference: Food Innovation and Engineering Conference

Year: 2019

Proceeding: 2019 Food Innovation and Engineering (FOODIE) Conference

Group: General Submissions

Session: Processing (cont)

Time: Tuesday, December 10, 2019 - 9:30am-9:45am

Authors:

Arikal, A. O., University of California, Davis
Scott, W. T. Jr., University of California, Davis
Block, D. E., University of California, Davis

Two key metabolic activities relevant to industrial wine fermentations are nutrient utilization efficiency and tolerance to high ethanol concentrations exhibited by commercial wine yeast strains. Therefore, studying the details of yeast metabolism is of great interest to develop ways to control stuck or sluggish fermentations. Using a computational method called dynamic flux balance analysis (dFBA) combined with a genome-scale metabolic model of yeast we managed to simulate growth kinetics of *S. cerevisiae* under enological conditions. Using a metabolic model also allowed us to predict flux (reaction rate) values of each metabolic reaction in the cell, which helped to visualize metabolic pathways that are important in nutrient utilization efficiency of*Saccharomyces cerevisiae*. However, despite being a robust way to study yeast metabolism, current genome-scale models do not accurately represent the biomass growth equation which results in lower predicted maximum cell concentrations compared to experimental values, indicating that it is essential to experimentally measure major cellular components. This way, variation in biomass composition between different yeast strains can be assessed and the resulting improved models used to gain better insight into metabolic differences between strains. In this study we carry out a sensitivity analysis that demonstrate how much the model is affected by changes in major cell components including DNA, RNA, total lipid, protein and carbohydrate in biomass equation of *Yeast 8.3.3* (developed by the Nielsen research group at Chalmers University). We also demonstrate that using measured values of the biomass components improves our computational predictions. We can also more accurately predict metabolic fluxes for various yeast strains and understand the variation in metabolism between different strains that leads to disparities in nutrient utilization efficiency.

Performance Digital Twin for the Food & Beverage Industry

Conference: Food Innovation and Engineering Conference

Year: 2019

Proceeding: 2019 Food Innovation and Engineering (FOODIE) Conference

Group: General Submissions

Session: Processing (cont)

Time: Tuesday, December 10, 2019 - 10:00am-10:15am

Authors:

Longwell, G., Siemens
Aglave, R., Siemens PLM Software
Kopcha, S., Siemens

Virtually every food and beverage company is talking about innovation and digitalization. Staying ahead in the innovation race requires leveraging the advantages of digitalization in each of the three steps of the value chain: Ideation, Realization and Utilization. A common challenge in the food industry is bringing new products to market by effectively scaling up the process. This requires paying attention to quality during production and ensuring process efficiency by optimizing operating conditions.

How do fluid flow and hydrodynamic behavior in food processing equipment affect the yield and throughput? Can we look at the fundamentals of processes in order to achieve macro level gains in efficiency and energy savings? Which of such unit operations can be investigated from fundamental perspective using simulations? Are there possibilities to utilize such methods for bottling and packaging as well? What are the extents of saving one can realize?

We will attempt to answer some of these questions in this presentation based on examples and cases studies.

From an engineering perspective, the topics covered will include stirred vessels, dispersing gas and solids in liquids, behavior of high viscosity fluids, fluid structure interactions in mixing vessels, and optimization of process equipment. The tools and methods we discuss are generally termed as the Performance Digital Twin. The key enabling technologies include Computational Fluid Dynamics (CFD) and particle modeling with Discrete Element Method (DEM).

Techno-Economic Analysis of a Plant-Based Intense Natural Sweetener Production for Reduced Global Sugar Consumption

FOODIE
FOOD INNOVATION AND ENGINEERING CONFERENCE

Conference: Food Innovation and Engineering Conference

Year: 2019

Proceeding: 2019 Food Innovation and Engineering (FOODIE) Conference

Group: General Submissions

Session: Processing (cont)

Time: Tuesday, December 10, 2019 - 9:45am-10:00am

Authors:

Kelada, K. D., University of California, Davis
Nguyen, R., University of California, Davis
Maborang, M. K., UC Davis
McDonald, K. A., University of California, Davis
Nandi, S., University of California, Davis

There is currently a worldwide trend to reduce sugar intake due to the various adverse health effects linked with the overconsumption of sugars. Chemically synthesized sweeteners (CSS) have been used as an alternative to nutritive sugars in numerous applications, however, their long-term effects on human health remain controversial. Therefore, there is a shift in consumer preference towards non-nutritive sweeteners from a natural source. Thaumatins is a class of intensely sweet proteins found in arils of the fruits of the West-African plant *Thaumatococcus danielli*. The current production method through aqueous extraction from this plant limits its supply while the demand is high. Despite successful recombinant expression of the protein in several organisms, no biotechnological production platform has yet been established. In this study, we present a detailed process simulation model for a large-scale, transgenic plant-based production facility for thaumatin II. The model was used to assess how critical process parameters (i.e. the method of plant growth (field or indoor), thaumatin II expression level in the plant tissue, and production capacity) affect certain economic parameters (total capital investment, annual operating cost, and cost of goods sold). For the base case scenario (50 MT thaumatin/year, 1.5 g thaumatin/kg fresh weight, and 80% downstream recovery), our analyses indicate that the total capital investment is $71.5 million dollars and the cost of goods sold is $356/kg including facility dependent costs. Considering a sweetness index range of x2,000-8,600 sweeter than sucrose, depending on the existing sucrose concentration, the application of thaumatin can have a tremendous positive impact on global health, as well as a potential reduced cost of production for sugar-sweetened products.

What Key Questions Should You be Asking Your Project Team during Scale-up of a Food Process?

FOODIE
FOOD INNOVATION AND ENGINEERING CONFERENCE

Conference: Food Innovation and Engineering Conference

Year: 2019

Proceeding: 2019 Food Innovation and Engineering (FOODIE) Conference

Group: General Submissions

Session: Processing (cont)

Time: Tuesday, December 10, 2019 - 10:15am-10:30am

Authors:

Hartford, C., Jenike & Johanson, Inc.

Ready to grow your production line? Good - but what challenges lie ahead specifically related to the flow of powders and ingredients through the new, larger equipment? Many issues in processing can arise as processes are scaled up for higher capacities. Often the science behind producing the new product is focused on and the practical aspect of the handling of the ingredients is pushed to the backburner.

To achieve the desired production throughput, quality and safety, it is critical that ingredients flow through the process without issues in the handling equipment. This presentation discusses best practices for scale-up production to minimize risk and achieve a reliable material handling system in food production. As an outcome, attendees will come away with a list of valuable questions they should be asking their project team so start-up occurs as planned, producing high quality products safely from the beginning.

Food & Ingredients for the Plant-Based Foods Market - the Seed to Solution Model

Conference: Food Innovation and Engineering Conference

Year: 2019

Proceeding: 2019 Food Innovation and Engineering (FOODIE) Conference

Group: General Submissions

Session: Food Architecture

Time: Tuesday, December 10, 2019 - 11:00am-11:25am

Authors:

Atchison, N., University of Minnesota

Plant-based foods have grown 31% over the past two years and are on pace to increase 18X by 2030 - to a market size of $85 billion. Food makers aiming to create plant-based foods that meet the taste, nutritional, and texture properties of their more conventional counterparts are faced with challenges working with new proteins and texturizers. Designing foods that meet Organic and Non-GMO Project standards increases this complexity. PURIS aims to solve these challenges of food formulators by providing innovative plant-based ingredients that provide neutral flavor, wide ranging functionality, and unique use cases for the most challenging formulation issues. Ingredients designed from the seed up to perform in a wide range of applications from high protein ready-to-drink beverages, to bakery, to confection. In this talk we will explore some of the challenges facing food formulators in the plant-based space and how ingredient suppliers are driving innovation to fuel the growth of the market.

Ensuring Safe Consumption of Frozen Foods: Challenges and Opportunities

FOODIE
FOOD INNOVATION AND ENGINEERING CONFERENCE

Conference: Food Innovation and Engineering Conference

Year: 2019

Proceeding: 2019 Food Innovation and Engineering (FOODIE) Conference

Group: General Submissions

Session: Sustainability & Security

Time: Tuesday, December 10, 2019 - 3:25pm-3:50pm

Authors:

Gummalla, S., American Frozen Food Institute

Frozen foods are at an important intersection in the food industryâ€™s evolution, especially from the perspective of improved manufacturing practices and meeting consumersâ€™ changing needs. Food safety risks associated with frozen foods have raised significant concerns, among food manufacturers, regulatory agencies, and consumers. Varied consumer practices, particularly in the way consumers follow cooking instructions and diverse new product uses and trends have only heightened interest.

This presentation will cover the gamut of challenges facing the industry, whether aging infrastructure, lack of modernized equipment, challenging regulatory policies or consumer behaviors. Indeed there are opportunities at each stage of the production cycle, innovation in processing and equipment and novel new antimicrobial technologies. The American Frozen Food Institute has developed a comprehensive food safety strategy to address these challenges, and will share how it has set in motion a program to combat food safety risks associated with frozen foods by addressing gaps in science, guiding practical regulatory policy, development of best manufacturing practices and training, investing in consumer research and education, and fostering innovation. AFFI hopes to deliver a compelling case to have companies and technical experts from varied disciplines to develop innovative solutions ensuring frozen foods are made safe for consumption in every stage of the farm to fork continuum.

Fat for the Future: Designing Multifunctional Molecular Oleogels As Trans Fat Alternative

FOODIE
FOOD INNOVATION AND ENGINEERING CONFERENCE

Conference: Food Innovation and Engineering Conference

Year: 2019

Proceeding: 2019 Food Innovation and Engineering (FOODIE) Conference

Group: General Submissions

Session: Sustainability & Security

Time: Tuesday, December 10, 2019 - 3:00pm-3:25pm

Authors:

John, G., The City College of New York,CUNY

Promotion of public health by delivering healthy food is a subject of research with national importance. Fats are among the most vital macronutrients in our diet as they could influence the organoleptic properties of the foods. New regulations have brought new challenges in structuring vegetable oils without any chemical modification. This led to the development of other alternatives and, among them, â€œoleogelsâ€, a potential strategy that could form bulk fats via physical structuring of vegetable oils by eliminating trans fats. The crystalline composition of oil-based formulations greatly impacts the aesthetic and textural properties of gel and cream formulations. The present research discusses using the self-assembly concept of nanotechnology to develop healthy and amphiphilic oil-structuring agents in food engineering and for personal care products. By conjugating active substrates to develop an array of molecular gelators they serve as programmed functional materials derived from biomass. Chosen precursors for oil gelling amphiphiles (medium chain fatty acids and non-reducing sugars) are FDA approved and GRAS materials; hence the amphiphiles are non-toxic and exhibited high cell viability at concentration 50 Î¼g/mL. Sugar-based gelators readily formed nanoscale lamellar structures to form a coherent network at very low concentrations (1-3% wt/v), which congeal a wide range of lipophilic oils. The efficiency was computed in terms of mechanical, thermal and structural properties and found to be a function of type and concentration of structuring agents. Select examples of biocatalytically enabled molecular gelators and their oleogels will be discussed. By developing structuring alternatives to chemically processed oil formulations of the future will not only be sustainable, but tailored by structure to function as needed. **Keywords**: Oleogels, Self-Assembly, Oil Structuring Agents, Trans fat Alternatives, Naturally Derived, Sustainable

Food System Decarbonization

FOODIE
FOOD INNOVATION AND
ENGINEERING CONFERENCE

Conference: Food Innovation and Engineering Conference

Year: 2019

Proceeding: 2019 Food Innovation and Engineering (FOODIE) Conference

Group: General Submissions

Session: Sustainability & Security

Time: Tuesday, December 10, 2019 - 3:50pm-4:15pm

Authors:

Mosey, G., National Renewable Energy Lab (NREL)

Food systems include the production, processing, packaging, transportation, storage, wholesale, retail, consumption, waste and disposal of food. Even more broadly the food systems touch chemical manufacturing, agriculture, food and beverage manufacturing, the service industry, and municipal waste. All these systems require energy as a key input, a significant portion of which is provided by fossil fuels, either through direct use in farming activities and food processing, or through indirect use in the production of agricultural inputs and machinery (Wood et. Al, 2010). While fossil fuels have played a significant role in increasing food productivity, it is not without costs. Exposure to fossil fuel price volatility and contribution to climate change can detrimentally affect food prices and increase the occurrence of crop stressors such as drought, respectively (Canning et al., 2010; and IPCC 2019). Fuel price volatility most acutely affects countries that depend on fossil fuel imports, while the effects of climate change jeopardize the food security and well-being of the global population. In the developed world, the use of energy in the food systems has grown steadily for some time. At the turn of the 20th century, roughly 40 percent of Americans were farmers; today that number is less than 1 percent (Lusk 2016). This immense increase in labor productivity is almost solely attributable to increased capital sophistication, driven entirely by fossil fuels. And as developing countries transition to middle income, their food systems will likely develop similarly to the developed world, vastly increasing global energy demand in the food systems. Addressing energy use in the food systems is therefore vital. Renewable energy and energy efficiency offer a solution to shield the food systems from volatile fossil fuel prices, decarbonize food production, and increase profitability.

AUTHOR INDEX

AIChE
120 Wall Street, FL 23
New York, NY 10005-4020

ISBN 978-1-7138-0998-2

9 781713 809982

CURRAN ASSOCIATES INC.
proceedings
.com